LIFE SKILLS
IN THE PACIFIC

Basic Livestock

Steven Potek

OXFORD

Contents

Introduction

Basic Livestock is about raising and caring for animals such as chickens, pigs, rabbits and fish as a source of nutritious food for the family. It is also about using livestock as a source of income by selling produce from these animals at the market. It contains instructions on how to raise animals, including construction of their housing, feeding requirements, information on diseases that affect specific animals, and how to make money by selling animals and animal products at the market.

This book is written to complement the newly introduced 'Making a Living' subject for upper primary students. It supports the philosophy of Education, which emphasises relevant education and self-reliance. It teaches students useful skills and relevant knowledge to enable them to become useful members of their local communities.

Strand: *Managing Resources*
Sub Strand: *Crop and Animal Management*
Outcomes: 7.1.3: *explain appropriate crop management and animal husbandry practices and demonstrate these through undertaking a practical project*
8.1.3: *plan, design and implement a crop or animal project suited to local conditions and resources aimed at generating an income*

I encourage all students to use the information in this book to take up an animal project, as it is an excellent means of providing a family income.

Stephen Potek

Chickens

Chickens are the most popular kind of poultry in Papua New Guinea. They thrive in the tropics and are an excellent source of body-building protein. Chickens can be raised for meat or eggs or both, and even their feathers have many uses.

Chickens are identified according to **class** and **breed**. There are four classes and about 200 different breeds. Native chickens move freely around villages and have adapted to the tough conditions they live in. They are generally smaller, grow slower and lay fewer eggs than **hybrid** chickens, which are crossbred from two different classes.

Chickens are easy to rear, cheap to feed, can survive local conditions and produce healthy meat and eggs. Chickens that are well cared for, with proper food and housing, will be healthier, fatter and lay more eggs than those left to fend for themselves around a village.

Village chickens

Starting a chicken project

Care and management of chickens is essential if your chicken project is to be successful. Good management usually means success and profit while poor management usually ends in failure and financial loss. A successful project means that you have managed all aspects of chicken production effectively and have produced healthy chickens for meat or eggs, which can then be sold at the market.

To manage chickens successfully:

- know your market
- select good healthy breeds
- provide good housing
- feed chickens properly for quick growth and good health
- control diseases and pests for good health
- keep financial records.

Market

Before deciding what type of chickens to raise, and whether to rear them for meat or eggs, you should find out about market conditions in your local area. If there is a demand for chickens and eggs, the price paid for them will be high; however, if there is an over-supply, the price will be low.

Chickens for sale at the market

Market *continued …*

The size of your project will also depend on how much money is available, what facilities you have, access to roads and markets, and your production knowledge and sales skills.

Eggs for sale at the market

Start by buying just a few chickens. One person will usually be able to provide enough food for a maximum of 16 chickens but extra help would be required to raise a larger number. Once they have grown to a good size, sell them at the market and use the money to buy more chickens. Never buy more chickens than you can manage properly.

Note: *Chickens can be sold live or* ***dressed****. (It is better to sell live because many farmers do not have electricity to refrigerate dressed chickens – the meat will spoil in a few hours and cannot be eaten.)*

Types of chickens

There are three types of chickens:

- Broilers (or fryers) are raised for their tender meat.
- Layers are raised for their eggs. (When old layers stop laying eggs they are slaughtered, but their meat is tough and needs to be boiled for quite a long time.)
- Dual-purpose chickens are raised for their eggs and meat. They are slaughtered for their meat when they are ten to 12 weeks old. If they are raised for eggs, they start laying at five to six months.

Very young chickens need a lot of work – they require special attention and care to stay healthy. When you are starting a chicken project, it's best to buy four to six-week-old chickens. They will be more expensive than very young chickens, but there is less chance of them dying at this age.

Housing

Good chicken housing will lead to better eggs and more profit. Chickens will lay regularly and eggs can be collected easily as they will not be scattered around the village. A chicken house should be light, well ventilated and easy to clean. It will give the chickens shelter from the weather and protection from other animals and thieves.

In most villages, chickens are kept inside all morning to lay. They are released at noon to hunt for food, fed some grain or seeds in the chicken house late in the afternoon, then locked up for the night.

Bush materials, such as sago palm leaves, kunai and bamboo, are cheap and can be used to make very good chicken houses.

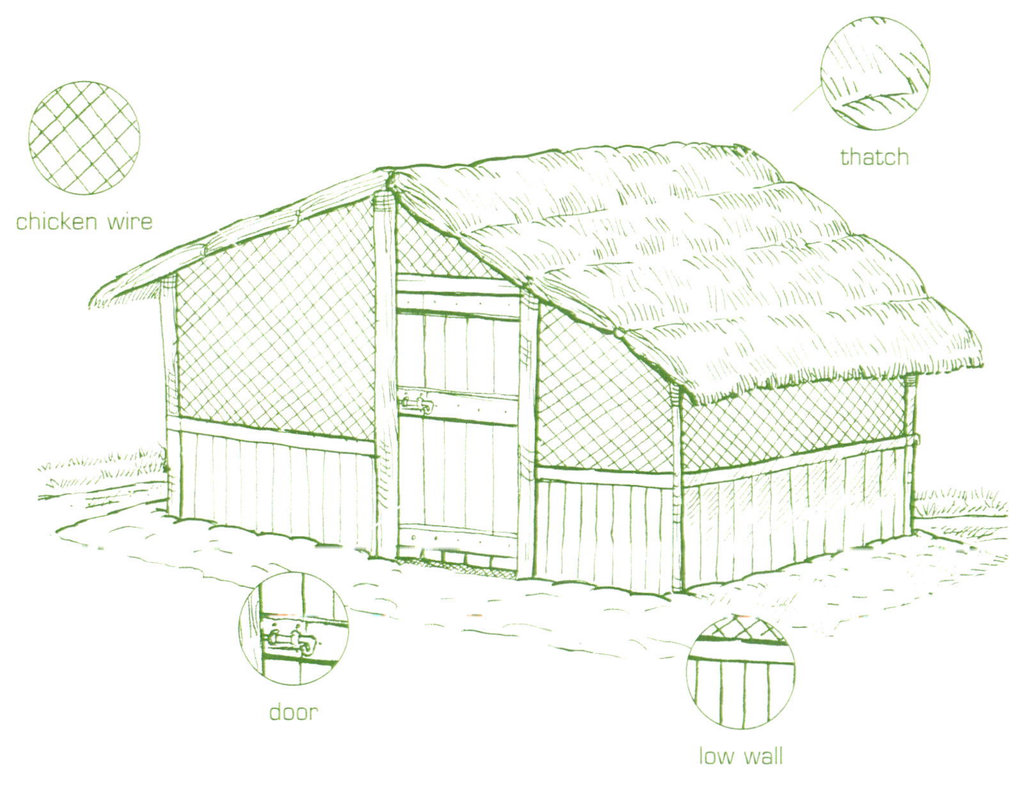

Housing *continued ...*

Inside the chicken house, you will need to provide suitable containers and boxes for feeding, drinking, **roosting** and nesting. These can be made cheaply from bamboo, timber and old 20-litre drums.

- You can cover the water troughs with wire to stop chickens from drowning.
- Water troughs can be attached above the floor to prevent spilling onto litter.

bamboo water trough

bamboo feed trough

Housing *continued ...*

- Feed troughs should only be half-filled to avoid wastage.

timber food trough

- Each nesting box should be about 40 by 50 centimetres – 20-litre drums make excellent nests.

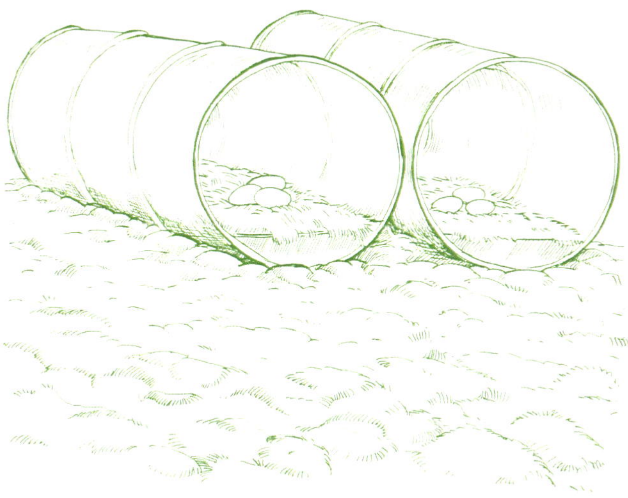

Housing *continued ...*

To prepare the chicken house for the chickens:

- Clean out any old litter and place it on the compost heap.
- Disinfect the floors and walls using detergent mixed with water.
- If you are in the highland provinces, make sure you have sacking to hang around the chicken house and help keep the birds warm at night.
- Wash, disinfect and place feeders and water troughs in position.
- Repair and replace any damaged troughs.
- Cover the floor with fresh dry litter, about 20 centimetres deep, using sawdust if possible.

Deep litter system

Deep litter is any dry material such as coffee husks, peanut husks, sawdust, dry leaves or chopped grass that is placed ten centimetres to 1.45 metres deep on the floor of the chicken house. The litter must be raked over once a week to stop it becoming hard and to mix the chicken droppings through it. It should also be changed regularly.

The litter must be kept dry so the roof of the chicken house must be waterproof, the overhang of the roof must be large enough to stop rain blowing in, and drinking water should not spill on the litter.

The chicken house should not be overcrowded – if there are too many chickens the litter becomes hard and diseases can spread.

The advantages of a deep litter system are:

- The litter acts as **insulation** – it stays cool in hot weather and warm in colder weather so chickens can burrow into the litter to keep cool or warm.
- Chickens scratch in the litter. This keeps them busy and stops them picking at their own feathers and eating their eggs. It also helps control lice.
- Deep litter does not smell, stays dry, and flies will not breed in it.
- Deep litter makes useful fertiliser for growing healthy crops.

Feeding

Feeding is by far the most important aspect of **poultry** management. No animal can grow and produce well unless it is fed properly. The production of meat and eggs is dependent upon the correct type and quantity of chicken feed. All feed must have the correct amounts of carbohydrates, fat, protein and vitamins and minerals. If only one of these things is lacking in the diet, then production of meat or eggs will fall.

Note: *To produce eggs, a different type of feed is required from that which is used to produce chickens for meat.*

Feeding *continued ...*

Many good foods for feeding poultry can be grown in Papua New Guinea:

- Energy foods like corn, sorghum and rice can be mixed with other high protein foods.
- Staple foods such as kaukau, yam and taro are also good energy foods but they must be cooked before they are fed to poultry.
- Protein foods like beans and peanuts are an excellent source of protein but they must be cooked before fed to poultry.
- Protective foods such as dark leafy vegetables should also be fed to the chickens so that they develop strong bones and strong eggshells.
- Commercial chicken feeds have all the **nutrients** for chicken growth but growing food for chickens can save a lot of money.

A selection of greens

Sacks of commercial chicken feed

Health

Disease control is an important part of chicken rearing. Diseases are caused by **bacteria** or **parasites** but can be avoided if chickens are cared for properly. Chickens that suffer from overcrowding, poor nutrition or lack of water can become weak, stressed and vulnerable to poor health. If there is an outbreak of infectious disease:

- separate the sick chickens from the healthy ones
- wash and disinfect the food and water troughs
- wash and disinfect the chicken house
- replace old litter with fresh dry litter.

Chicken deaths are also caused by starvation, thirst or injuries so:

- provide clean drinking water at all times
- avoid overcrowding
- provide enough feeding troughs and nests
- ensure the chickens have enough feed and a balanced diet.

Financial management

A budget is a statement of:

- how much money will be spent on what
- how much income is expected and when.

A chicken project budget should include:

- what types of chickens are to be reared
- what resources are required to rear chickens
- what and how much will be sold
- what will be the income from the sale.

***Note:** If the income (money from sales) is higher than the cost (expenses) you will make a profit.*

Financial management *continued ...*

Planning a budget

Budgeting is part of planning a chicken project. You need to collect information such as the price of chickens, chicken feed and building supplies so you can start to prepare a budget. Keep records in writing so you will know exactly how much money was spent to rear the chickens, how much money was made selling them or their by-products, and whether you made a profit.

It is important to learn these skills and keep records in writing. Written records will show exactly:

- how much money was made selling chickens and chicken by-products
- how much it cost to keep chickens and produce chicken by-products
- whether the money received was more money than the money spent.

To be a successful chicken farmer and run a profit-making business, no matter how small, you need to keep an accurate journal and financial records. The journal should include growth and production information, which is basically everything that happened while rearing the chickens. Financial records should include all matters relating to expenses and revenue. It is important to keep records of how much it costs to rear your chickens and how much revenue you receive for it. It is best to keep a simple financial record for every chicken or by-product produced – live chickens, eggs, feathers.

Sample farming journal

March 2007	
Monday	*Repaired chicken house roof* *Collected eggs*
Tuesday	*Cleaned out deep litter* *Re-filled water troughs* *Collected eggs*
Wednesday	*Repaired chicken house door* *Re-filled water troughs* *Collected eggs*
Thursday	*Sold eggs at market*
Friday	*Slaughtered two chickens* *Prepared chicken feathers*
Saturday	*Sold 5 live chickens at market*

Financial management *continued ...*

Sample financial record

Production costs

Date	Item purchased	Cost
March 2	40kg starter food	K55
March 14	40kg grower food	K50
March 15	40kg finisher food	K54

March 2007: Total costs K159

Revenue

Date	Chicken product sold	Value
March 10	8 live chickens (6 weeks old)	K160
March 25	45 eggs	K28

March 2007: Total revenue: K188

Summary: March 2007

Total revenue	K188
Total costs	K159
Profit/loss	**K29 profit**

Pigs

Pigs are very important in traditional Papua New Guinean society. In the highlands, people place a very high value on their pigs, which are used for ceremonial feasts, bride price, exchange and prestige.

Pigs grow and mature quickly, breed easily, are almost disease-free and can survive in very rough conditions including drought. They can also provide a cash income for rural families.

The following pig products can generate an income:

- pig meat (pork)
- live piglets
- pig **manure** for gardens
- tusks (front teeth) for making necklaces
- hairs for making brushes
- skulls for decoration and family wealth.

In most villages, pigs are left to wander around looking for food. They are sometimes given cooking scraps, dried coconut flesh or unwanted cassava or kaukau that has been boiled up for them. Pigs that are left to roam around freely do not need much care but they do not grow as well as pigs that are kept in pig houses.

Sacks of commercial pig feed

Starting a pig project

Care and management of pigs is essential if your pig project is to be successful. Good management usually means success and profit while poor management usually ends in failure and financial loss. A successful project means that you have managed all aspects of pig production effectively and have produced healthy pigs for breeding, live pig sales, or meat that can be sold at the market.

To manage pigs successfully:

- know your market
- select good healthy breeds
- provide good housing
- feed pigs properly for quick growth and good health
- control diseases and pests for good health
- keep financial records.

Market

Before deciding what kind of pigs to raise, and whether to rear pigs for meat or live sale, you should find out about market conditions in your local area. If there is a demand for pigs or pork, the price paid for them will be high; however, if there is an over-supply, the price will be low.

The size of your project will also depend on how much money is available, what facilities you have, access to roads and markets, and your production knowledge and sales skills.

Never keep more pigs than you can manage properly. Start slowly, perhaps with a small co-operative, sharing breeding stock to produce healthy piglets. A payment can be made to the owner of the breeding **boar** for mating with each individual **sow**. Once piglets are born, sell a few to pay for production costs and keep the rest for breeding or grow them to a good size for meat production.

Types of pigs

In Papua New Guinea, there are three kinds of pigs that are most common:

- Village (bush) pigs are good at grazing and eating raw kaukau because they have learnt to forage for themselves. They do not like living in small spaces and will grow fast if they are well fed.
- Imported large white pigs like to live in small houses and eat cooked kaukau. Village pigs bred with white pigs will produce strong, healthy, fast-growing piglets.
- Imported Tamworth pigs.

The life cycle of a pig

The **gestation** period for a pregnant sow is about 16 weeks. A pregnant sow should be well fed with protein-rich foods and young greens. During pregnancy, the sow increases in weight by 40 to 45 kilograms. The amount of feed should be gradually increased to supply the nutrients required for this increase in size.

A sow and her piglets

Types of pigs *continued ...*

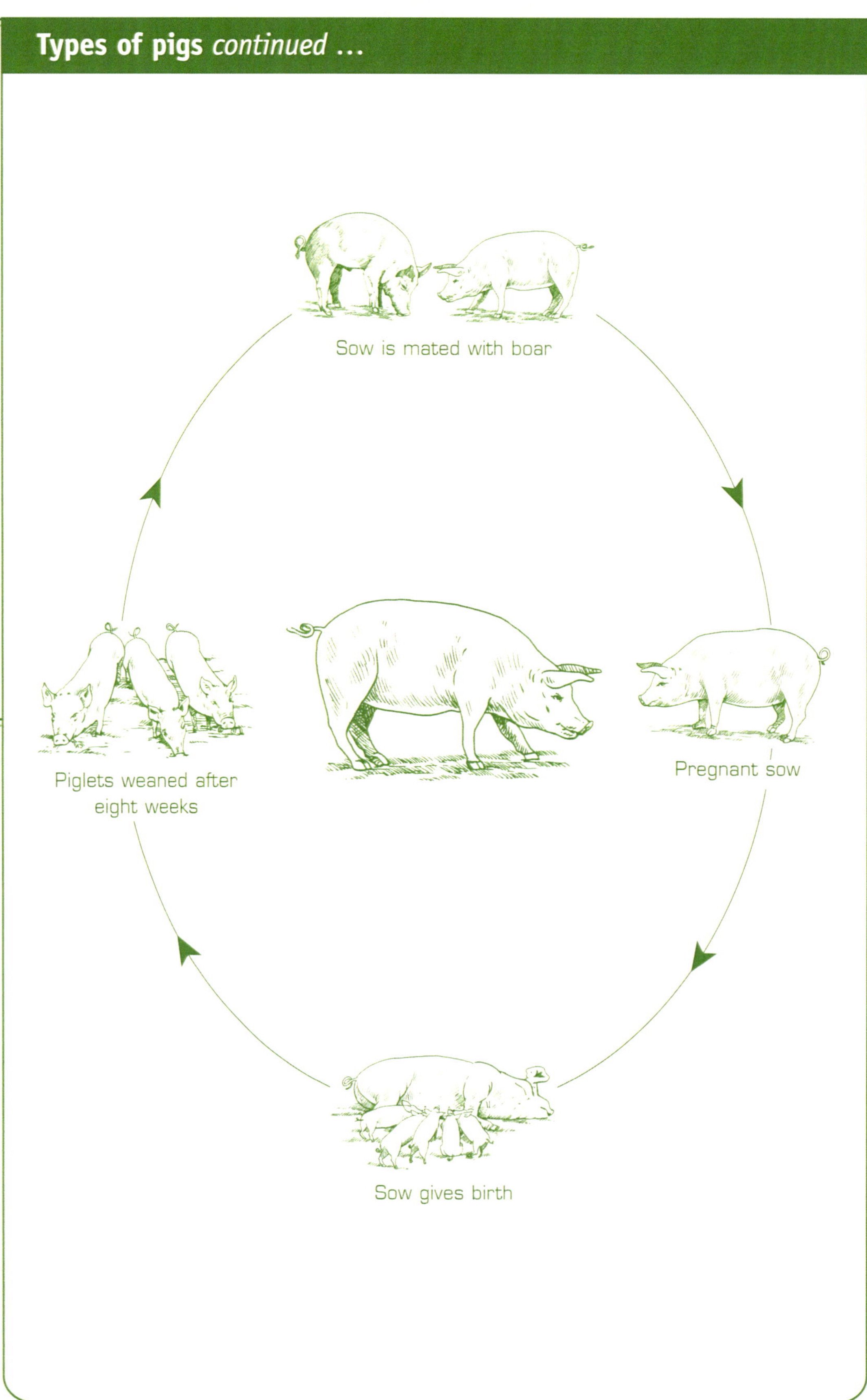

Housing

Make sure your pigs have a clean, dry, sheltered environment. Permanent pig houses enable many pigs to be kept in a relatively small space and prevent the spread of disease. A pig house should have a roof to keep pigs out of the rain and sun, and good **drainage** – small drains should be dug around the house to stop water coming in. The pig house should also be close to a water supply, such as a well, stream or tap, so that they will have fresh drinking water close by – otherwise it must be carried to them quite often. It should also be well ventilated to prevent the air becoming stagnant, warm and humid. (A pig that suffers heat stagnation will lose its appetite, lose weight and is likely to catch diseases.) In all pig houses there must be one drum of fresh water and another drum for feed.

If pigs are grazing in kaukau fields, then fences should be built around the grazing area and there must be enough kaukau for them to feed all year. They will still need some basic housing to provide protection from the weather and a place to sleep.

Housing *continued ...*

Local conditions will determine the best type of pig house. On the coast, a concrete floor or a slatted wooden floor is best. In the highlands and lowlands, a concrete floor or deep litter floor is best. Bush materials such as kunai grass, bush timber and bamboo can be used to build simple cheap houses but maintenance will be required.

Types of pig houses

The slatted floor house

This is a very good house for coastal areas. Pig **dung** falls through the slats and only needs to be cleaned from under the floors every seven to ten days.

sago palm or kunai grass roof

dung

slats

Housing *continued ...*

The concrete floor house

This house is good for both coastal and highland areas. It is simple to build, but sand and cement are needed to make the concrete floor. The floor should be built so that it is higher on one side than the other. The pigs lie on the high side and dung will fall on the lower side. (Pigs do not like to lie in their own dung.)

Housing *continued …*

The deep litter house

This house is very good for the highlands and requires little maintenance. The house is built in the same way as a slatted house but with a layer of stones on the floor. Put fresh kunai grass or coffee husks over the top of the stones every day. The pigs can lie in the deep litter when it is cold and this will help to keep them warm.

Housing *continued ...*

The movable shelter

A simple movable shelter is good for pigs that are grazing in kaukau fields. The house needs a strong door – two strong door posts and four pieces of wood act as guides so that slats can be put in place. The slats slide in through the top of the guides and can be removed the same way to let the pigs out.

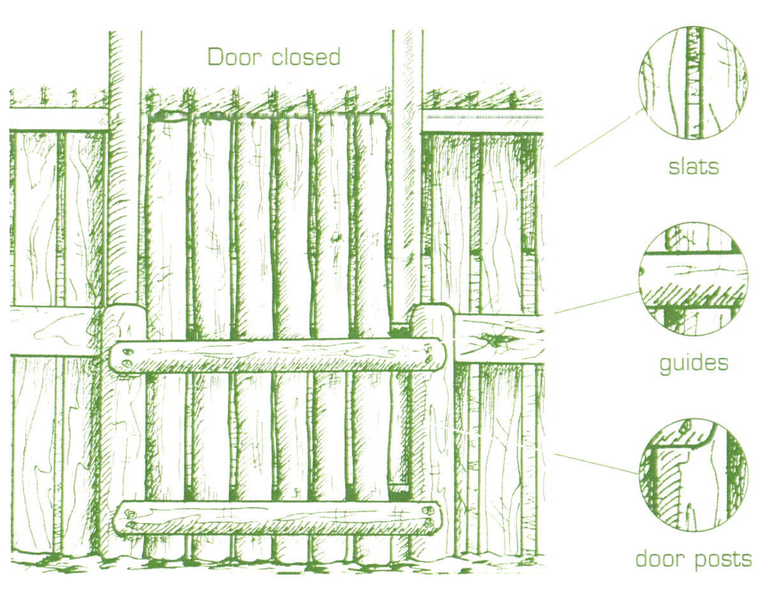

Feeding

Pigs need all the following foods:

- Protein foods – commercial feed/ concentrate, soya beans (cooked), fishmeal, winged beans and peanuts (cooked).

- Energy foods – starchy foods such as kaukau, taro, yams, copra, coconut, tapioca (fresh without greens), banana and grains.

Feeding *continued ...*

- Vitamins and minerals – dark green leafy foods such as kaukau leaves, fishmeal, concentrate and cod liver oil.
- Fresh clean water.

Whether you have one pig or many, you need to feed it well so that it can grow well, stay healthy and produce a good litter. Pigs need a balanced diet of energy, protein and protective foods. Allow pigs to eat as much energy food as they want – they will grow and be ready for sale more quickly. Vitamins are also necessary for good health because pigs often lack vitamins A, B12 and D.

Pigs of different sizes and ages have different needs. Big pigs need more energy food and sows should be given double the quantity of meal, green leaf and other feed that is given to male or young pigs. Small or weak pigs should be separated to ensure they get their share of food. Food should be put in containers or on clean ground if the food is dry enough. Pigs should not be able to lie in the food container and / or tip it over.

Feeding *continued ...*

Make sure your pigs have clean drinking water. Pigs that get fresh and cooked feeds require less water than those getting dry meal but still need between two and four litres of water each day, depending on the size of the pig. Sows feeding piglets on milk require at least ten litres of water per day. Pigs should not be able to lie in the water container or tip it over.

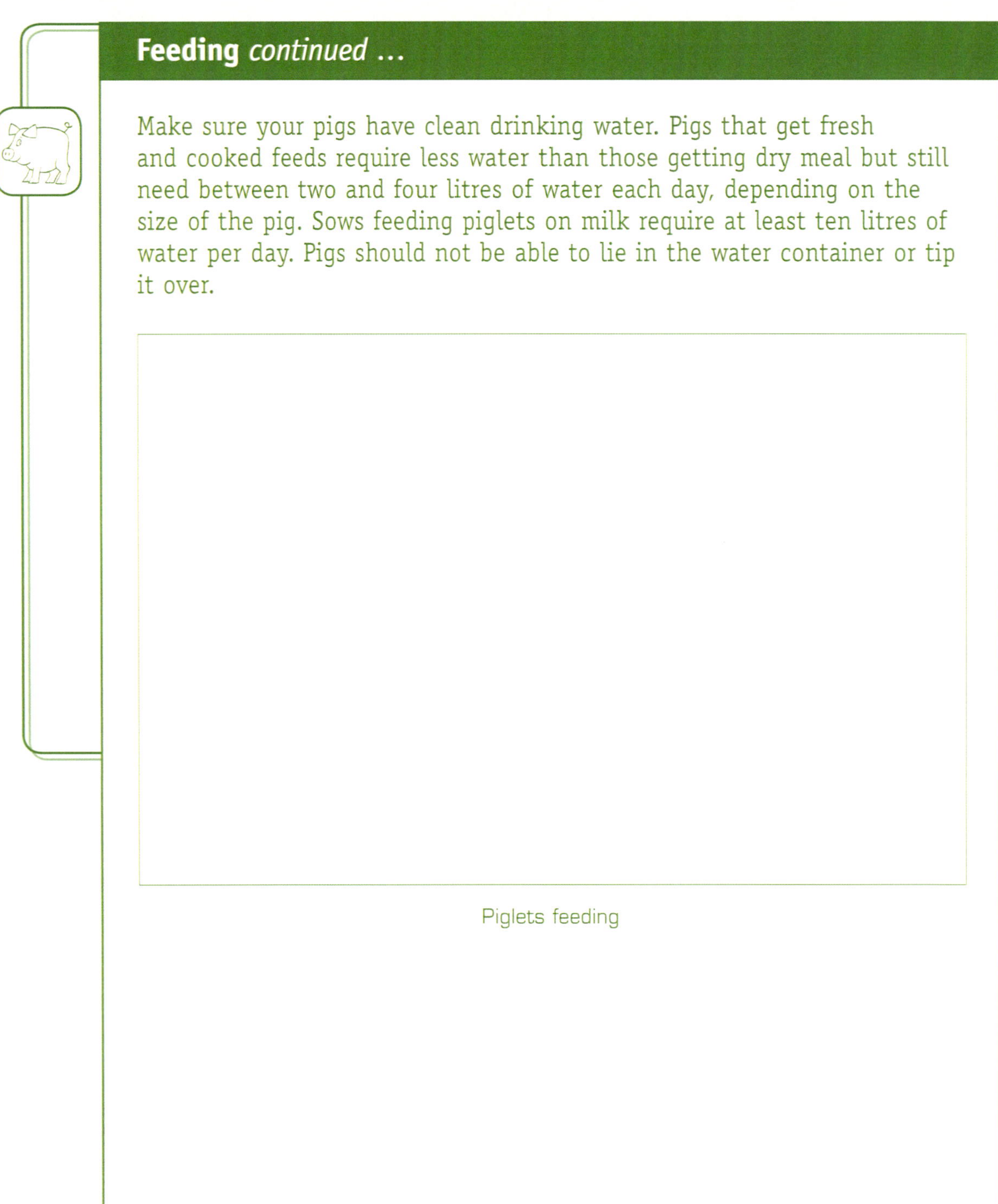

Piglets feeding

Health

Healthy pigs should not have pot bellies, rough coats or dry scaly skin. They should be active outdoors, digging and seeking food. Pigs that are not getting enough food and water, are exposed to too much sun, or have too many worms in their bodies will be unhealthy. Pigs that spend all their time scratching or rubbing against solid objects may have external parasites or **mites**. They can suffer from diseases caused by viruses, bacteria or worms, and mineral deficiencies including iron, calcium, phosphorus and salt. These problems can be avoided if pigs are housed and fed properly. If your pigs do not look well and are not gaining weight, seek assistance from a Livestock Officer.

If a pig doesn't have a balanced diet, the following may occur:

- Lack of vitamin A – pigs will grow slowly, lose their appetite, show signs of night blindness and lameness, and produce dead or weak piglets.
- Lack of vitamin B12 – pigs will grow slowly, lose their appetite, and suffer from anaemia.
- Lack of vitamin D – pigs will grow slowly and have weak bones. Some sunlight can help provide vitamin D.

A boar

Health *continued ...*

Choose good boars and use healthy sows, with seven pairs of evenly spaced teats, for breeding. Remove old or unproductive sows from the herd. One boar can service 15 to 20 sows.

The pregnancy lasts 114 days and a sow will be ready to mate again about one week after her **litter** has been weaned. A sow will usually not have enough milk to feed any more than seven or eight piglets.

Healthy baby pigs are weaned at five to six weeks of age. Do not let the pigs suckle too long or the sow will get too thin and may have difficulty regaining weight and getting pregnant again. Make sure baby pigs eat well before weaning by having fresh food available from the second week. They should always have plenty of clean drinking water.

Financial management

A budget is a statement of:

- how much money will be spent on what
- how much income is expected and when.

A pig project budget should include:

- what types of pigs are to be reared
- what resources are required to rear pigs
- what and how much will be sold
- what will be the income from the sale.

Note: *If the income (money from sales) is higher than the cost (expenses) you will make a profit.*

Financial management *continued …*

Planning a budget

Budgeting is part of planning a pig project. You need to collect information such as the price of pigs, pig feed and building supplies so you can start to prepare a budget. Keep records in writing so you will know exactly how much money was spent to rear the pigs, how much money was made selling them or their by-products, and whether you made a profit.

It is important to learn these skills and keep records in writing. Written records will show exactly:

- how much money was made selling pigs and pig by-products
- how much it cost to keep pigs and produce pig by-products
- whether the money received was more money than the money spent.

To be a successful pig farmer and run a profit-making business, no matter how small, you need to keep an accurate journal and financial records. The production journal should include growth and production information, which is basically everything that happened while rearing the pigs. Financial records should include all matters relating to expenses and revenue. It is important to keep records of how much it costs to rear your pigs and how much revenue you receive for it. It is best to keep a simple financial record for every pig or by-product produced.

Sample pig project journal

March 2007	
Monday	*Raked and replaced floor litter* *Re-filled water containers* *Prepared pig feed – cooked kaukau*
Tuesday	*Cleaned feed troughs* *Re-filled water containers*
Wednesday	*Checked pigs for skin diseases* *Re-filled water containers* *Checked the roof of the pig house for leaks*
Thursday	*Re-filled water containers* *Put pregnant sow in separate pen*
Friday	*Slaughtered one pig for meat* *Re-filled water containers*

Financial management *continued …*

Sample financial record

Production costs

Date	Item purchased	Cost
March 4	Bag pig meal concentrate	K200
March 11	Bag pig grower	K155

March 2007: Total costs K355

Revenue

Date	Pig product sold	Value
March 6	2 live piglets	K200
March 18	20 pieces smoked pork	K200

March 2007: Total revenue: K400

Summary: March 2007

Total revenue	K400
Total costs	K355
Profit/loss	**K45 profit**

Rabbits

Rabbit farming is becoming popular in Papua New Guinea. Rabbits breed and grow quickly, provide excellent meat and are cheaper to keep than many other animals. Rabbits can be used to provide meat, skins for pelts, furs for handicrafts and manure for gardens.

Starting a rabbit project

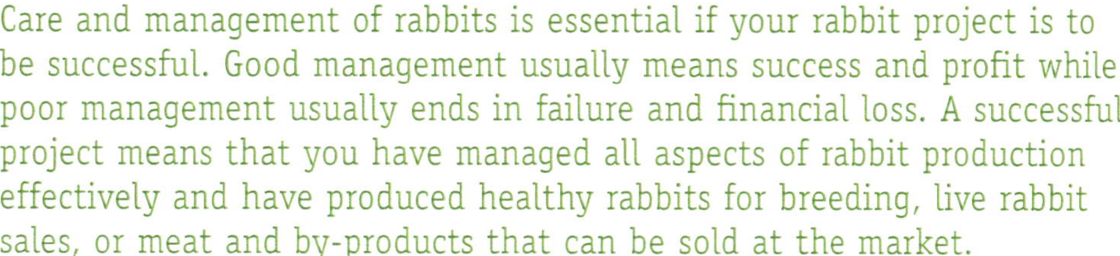

Care and management of rabbits is essential if your rabbit project is to be successful. Good management usually means success and profit while poor management usually ends in failure and financial loss. A successful project means that you have managed all aspects of rabbit production effectively and have produced healthy rabbits for breeding, live rabbit sales, or meat and by-products that can be sold at the market.

To manage rabbits successfully:

- know your market
- select good healthy breeds
- provide good housing
- feed rabbits properly for quick growth and good health
- control diseases and pests for good health
- keep financial records.

Market

Before deciding what kind of rabbits to rear, and whether to rear them for breeding, live sales or meat and by-products, you should find out about market conditions in your local area. If there is a demand for rabbits, the price paid for them will be high; however, if there is an over-supply, the price will be low.

The size of your project will also depend on how much money is available, what facilities you have, access to roads and markets, and your production knowledge and sales skills. Never keep more rabbits than you can manage properly.

Types of rabbits

Rabbit breeds available in Papua New Guinea are:

- New Zealand White rabbits – suitable for the highlands.
- Canberra Half Lop – suitable for both coastal areas and the highlands.

To breed rabbits, take the **doe** to the **buck** in his own cage. Leave them for two days and then remove the female. Return the female to the male after 12 days. If she rejects the male, this indicates that she is pregnant. Ensure the female has a box in which to make a nest to ensure a good survival rate for the babies.

Generally five to seven baby rabbits are born 30 to 33 days after mating. As long as they are kept out of the wind and rain, most will survive and can be weaned at five to six weeks of age. Remove weaned rabbits from the mother and she will be ready to mate again in about two weeks. A doe generally breeds well up to three years of age.

Baby rabbits grow quickly and weigh about two kilograms by three months. This is the best time to slaughter rabbits for meat as it is very tender and healthy. A two-kilogram rabbit will normally produce one kilogram of meat.

Housing

Rabbits need shelter as they will die if exposed to heavy rain. They also need to be protected from predators such as dogs, rats and snakes so it is best to build rabbit cages off the ground, on poles, with secure doors and no holes in wire netting.

The size of the cage depends on the number of rabbits. One rabbit needs a cage about one cubic metre in size. A cage for a female and babies should be slightly larger – about one and a half metres wide and long. The sides of the cage can be made of strong wire (which can be purchased from hardware stores) or bush materials such as cane, bamboo, split timber and black palm fronds.

A rabbit hutch with wire floor and deep litter tray

Housing *continued ...*

It is best to make the floor of the cage from wire, because this allows the manure to fall to the ground, or place a deep litter tray under the wire floor. The tray should be cleaned once a week and the rabbit manure can be used as fertiliser.

You need to ensure that:

- rabbit cages are fitted with water and food containers
- rest areas in cages are cleaned regularly
- nesting boxes for young rabbits are cleaned regularly
- litter is cleaned and replaced regularly
- cages are secure and the roof and walls have no holes.

Feeding

Rabbits need all the following foods:

- Greens – aibika, banana leaves, leucaena leaves, sweet potato vines, fresh green grass and many garden weeds.
- Energy foods – banana, raw kaukau, corn, coconut and copra meal and cassava.

These foods are all readily available throughout Papua New Guinea. Rabbits fed on these basic foods will be very healthy.

A selection of foods that are suitable for rabbits

Feeding *continued ...*

Rabbits need a balanced diet of energy foods and greens so that they grow quickly and gain weight. Rabbits that are fed only on greens may not gain weight and may have difficulty in breeding. Commercial rabbit pellets are a total rabbit food and rabbits fed on these will require nothing else except for a few greens; however, pellets cost money and can be expensive in the more remote areas of Papua New Guinea.

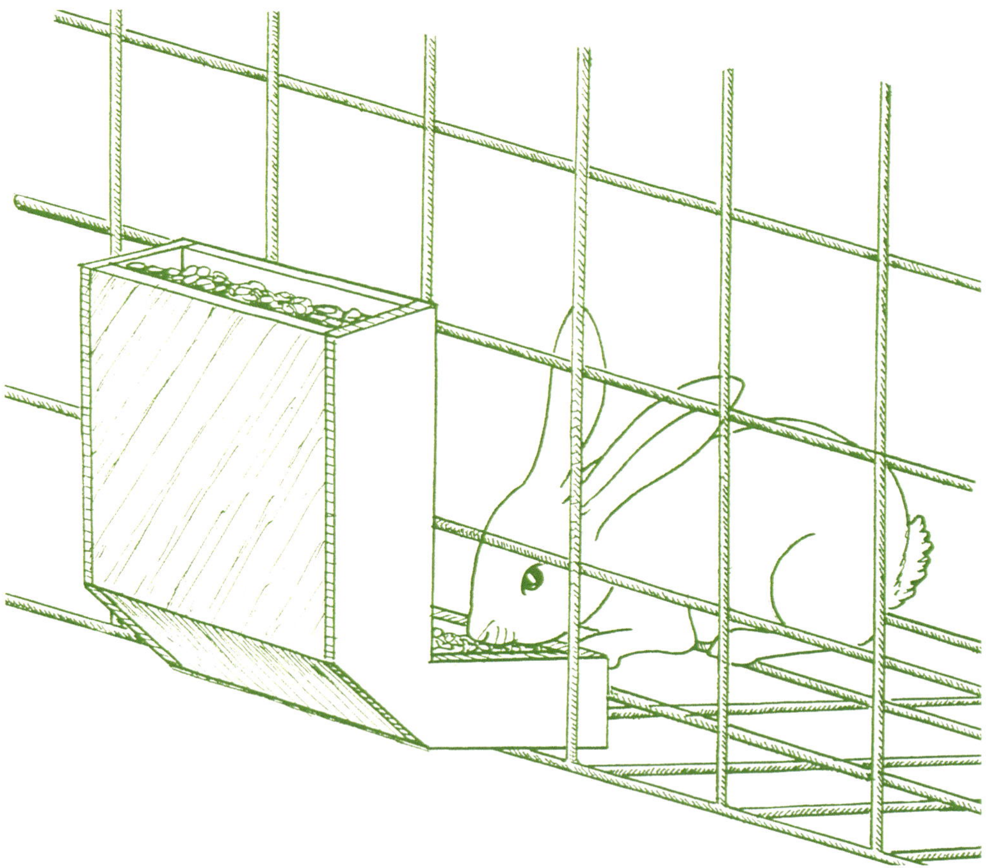

A rabbit eating pellets from a feeder

Health

Most rabbit diseases are the result of poor feeding (such as mouldy kaukau), a lack of energy foods, and dirty or wet cage floors that can cause sores and mites.

Financial management

A budget is a statement of:

- how much money will be spent on what
- how much income is expected and when.

A rabbit project budget should include:

- what types of rabbits are to be reared
- what resources are required to rear rabbits
- what and how much will be sold
- what will be the income from the sale.

Note: *If the income (money from sales) is higher than the cost (expenses) you will make a profit.*

Financial management *continued …*

Planning a budget

Budgeting is part of planning a rabbit project. You need to collect information such as the price of rabbits, rabbit feed and building supplies so you can start to prepare a budget. Keep records in writing so you will know exactly how much money was spent to rear the rabbits, how much money was made selling them or their by-products, and whether you made a profit.

It is important to learn these skills and keep records in writing. Written records will show exactly:

- how much money was made selling rabbits and rabbits by-products
- how much it cost to keep rabbits and produce rabbit by-products
- whether the money received was more money than the money spent.

To be a successful rabbit farmer and run a profit-making business, no matter how small, you need to keep an accurate journal and financial records. The production journal should include growth and production information, which is basically everything that happened while rearing the rabbits.

Financial records should include all matters relating to expenses and revenue. It is important to keep records of how much it costs to rear your rabbits and how much revenue you receive for it. It is best to keep a simple financial record for every rabbit or by-product produced – live rabbits, fur or meat.

Sample rabbit project journal

March 2007	
Monday	*Cleaned litter* *Re-filled water containers*
Tuesday	*Cleaned feed containers* *Re-filled water containers*
Wednesday	*Re-filled water containers* *Checked the cage for holes*
Thursday	*Re-filled water containers* *Prepared nest for pregnant female*
Friday	*Slaughtered one rabbit for meat* *Re-filled water containers*

Financial management *continued ...*

Sample financial record

Production costs

Date	Item purchased	Cost
March 4	Rabbits (1 male, 3 female)	K120
March 10	Rabbit feed	K50
March 15	Materials to repair cages	K100

March 2007: Total costs K270

Revenue

Date	Rabbit product sold	Value
March 6	Nothing sold	Nil

March 2007: Total revenue Nil

Summary: March 2007

Total revenue	Nil
Total costs	K270
Profit/loss	K270 (loss)

Fish

Fish farming in ponds is becoming popular in the highlands of Papua New Guinea, especially in areas where people cannot get fish easily from rivers or the sea. Fish is an excellent source of body-building protein and can also be used as food for livestock. Fish is the cheapest source of protein in Papua New Guinea because production costs are very low. It can also be a good source of income. Fish are sold fresh or smoked when they reach market size at about six to 12 months old. They can also be sold as **fingerlings** or adult fish for breeding, or used as fishmeal.

Starting a fish project

Care and management of fish is essential if your fish project is to be successful. Good management usually means success and profit while poor management usually ends in failure and financial loss. A successful project means that you have managed all aspects of fish production effectively and have produced healthy fish for breeding or meat that can be sold at the market.

To manage fish successfully:

- know your market
- select good healthy breeds
- provide a good pond environment
- feed fish properly for quick growth and good health
- control diseases and pests for good health
- keep financial records.

Market

Before deciding what kind of fish to farm, you should find out about market conditions in your local area. If there is a demand for fish, the price paid for them will be high; however, if there is an over-supply, the price will be low. Price is also determined by the size of the fish.

The size of your project will also depend on how much money is available, what facilities you have, access to roads and markets, and your production knowledge and sales skills. The management of a pond is very important and only suitable fish should be placed in the pond.

Fish for sale at the market

Types of fish

The best types of fish for pond farming in Papua New Guinea are carp and tilapia – both are freshwater fish. They breed well under most conditions and grow quickly to allow for quick harvesting of fish.

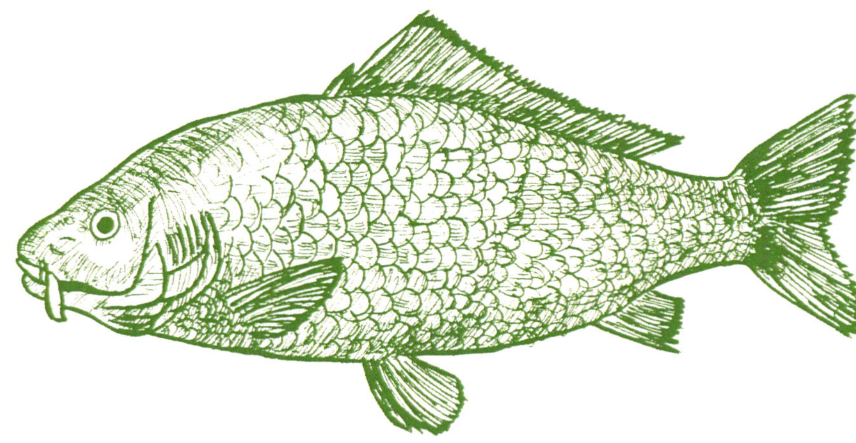

Carp

Tilapia

You can get freshwater fish from the nearest fisheries station of the Department of Agriculture and Livestock in your town or province.

Housing

It is best to build your pond on a low flat area at the bottom of a slope. Clay soil is essential to hold water. Dig a pond no more than two metres deep and include a drain with a valve that will allow water to run in and out of the pond. A drain is essential for pond cleaning and maintenance. You might need to start a small co-operative or organise a school group to help dig and build the pond.

When the pond is ready, allow fresh water to flow in. Plant some trees for shade, and water lilies and green grass around the pond for the fish to feed on. When the pond is full, the fish can be introduced into the pond. Do not allow water to flow out of the pond as the water must be kept still.

Feeding

Fish eat a lot of food each day and need sufficient food so that they can grow fast. Fish foods include:

- Food thrown into the water – cooked kaukau, cooked rice, leaves, grass or cooked yams.
- Small plants growing in and around the pond – throw **fertiliser** into the pond to allow small plants to grow quickly.
- Stock feed – commercial fish food is also available but usage may depend on your budget.

Financial management

A budget is a statement of:

- how much money will be spent on what
- how much income is expected and when.

A fish project budget should include:

- what types of fish are to be farmed
- what resources are required to farm fish
- what and how much will be sold
- what will be the income from the sale.

Note: *If the income (money from sales) is higher than the cost (expenses) you will make a profit.*

Planning a budget

Budgeting is part of planning a fish project. You need to collect information such as the price of fresh fish and smoked fish so you can start to prepare a budget. Keep records in writing so you will know exactly how much money was spent to rear the fish, how much money was made selling them, and whether you made a profit.

It is important to learn these skills and keep records in writing. Written records will show exactly:

- how much money was made selling fish
- how much it cost to farm fish
- whether the money received was more money than the money spent.

To be a successful fish farmer and run a profit-making business, no matter how small, you need to keep an accurate journal and financial records. The production journal should include growth and production information, which is basically everything that happened while rearing the fish. Financial records should include all matters relating to expenses and revenue. It is important to keep records of how much it costs to rear your fish and how much revenue you receive for it.

Financial management *continued …*

Sample fish project journal

March 2007	
Monday	*Fed fish* *Removed excess weeds* *Added fingerlings to the pond*
Tuesday	*Fed fish* *Added fertiliser to the pond*
Wednesday	*Fed fish* *Caught 5 fish for smoking*
Thursday	*Fed fish* *Drained some pond water* *Added fingerlings to the pond*
Friday	*Added fresh water to pond* *Fed fish*

Sample financial record

Production costs

Date	Item purchased	Cost
March 2	50 fingerlings	K100
March 6	3 sacks of stock feed	K150
March 9	Materials to build pond	K400
March 10	Wire cages and fertiliser	K500

March 2007: Total costs K1150

Financial management *continued …*

Revenue

Date	Fish product sold	Value
March 18	Adult fish (live weight)	K300
March 22	Young fingerlings	K200
March 26	Fishmeal	K500
March 30	Fish export	K800

March 2007: Total revenue: K1800

Summary: March 2007

Total revenue	K1800
Total costs	K1150
Profit/loss	**K650 profit**

low wall

Glossary

bacteria (singular: bacterium)	micro-organisms that can cause plant and animal diseases
boar	a mature male pig ready for mating
breed	a group of animals that look the same
buck	a mature male rabbit ready for mating
class	four main types of chicken that can then be grouped into different breeds
doe	a mature female rabbit ready for mating
drainage	removal of excess water from a piece of land
dressed	cleaned of feathers and slaughtered
dung	animal droppings from chickens, pigs, rabbits or cattle added to a garden to enrich the soil
fertiliser	substances that can be mixed with soil or water to add nutrients
fingerlings	very young newly hatched fish
gestation	the breeding cycle of a farm animal
hybrid	an animal that is crossbred from two animals of different breeds
insulation	materials that stay warm in cool temperatures and stay cool in hot temperatures
litter	a group of baby pigs
manure	animal dung from chickens, pigs, rabbits or cattle added to a garden to enrich the soil

mites	tiny animals that live off other animals, usually in the skin or body hair, and can cause sores or diseases
nutrients	any element that an animal, plant or human takes in and uses to grow such as nitrogen, potassium and phosphorus
parasites	plants or animals that live off other plants and animals, sometimes causing sickness or hindering growth
poultry	farmed birds including chickens, geese and ducks
roosting	perching inside a chicken house
sow	a mature female pig ready for mating

Notes